Ellison Alcântara
José Mendes

Biopolymer for thermal insulation

Ellison Alcântara
José Mendes

Biopolymer for thermal insulation

Development of a cassava starch biopolymer for thermal insulation

Imprint

Any brand names and product names mentioned in this book are subject to trademark, brand or patent protection and are trademarks or registered trademarks of their respective holders. The use of brand names, product names, common names, trade names, product descriptions etc. even without a particular marking in this work is in no way to be construed to mean that such names may be regarded as unrestricted in respect of trademark and brand protection legislation and could thus be used by anyone.

Cover image: www.ingimage.com

This book is a translation from the original published under ISBN 978-613-9-64337-0.

Publisher:
Sciencia Scripts
is a trademark of
Dodo Books Indian Ocean Ltd. and OmniScriptum S.R.L publishing group

120 High Road, East Finchley, London, N2 9ED, United Kingdom
Str. Armeneasca 28/1, office 1, Chisinau MD-2012, Republic of Moldova, Europe
Printed at: see last page
ISBN: 978-620-7-75835-7

ACKNOWLEDGEMENTS

Firstly, I thank God for blessing me with health, wisdom, strength and a lot of faith to reach this long-awaited goal.

To my supervisor Prof Dr José Ubiragi de Lima Mendes, for the opportunity and willingness to guide me, passing on knowledge and determination. He was always patient during the course of my work.

To my parents for their trust and help throughout my academic life, always encouraging and helping.

To Dayane Damásio for always encouraging and helping me.

To the fluid mechanics laboratory at UFRN, for providing the space and equipment to carry out the research.

To the secretary Luiz Henrique for his support and dedication to me and the students of PPGEM-UFRN.

To the evaluating professors for their contributions to this research.

SUMMARY

Environmental concerns have been growing in recent years, mainly due to the use of various petroleum-based products, which take years to degrade in nature. To minimise this problem, several studies have focused on the use of biodegradable natural materials. In this context, it was found that cassava starch can be used as a raw material to produce a biopolymer that can be used as a thermal insulator. The biopolymer was produced using a controlled thermal expansion process with volume fractions of 30 per cent water and 70 per cent cassava starch. The material's physical and thermal properties were analysed in order to obtain data on the viability of its use as a thermal insulator. The material had a thermal conductivity of 0.152 W/mK and could be used as an insulator in hot systems of up to 270 °C and a temperature resistance limit of 270 °C observed by thermogravimetric analysis. The cassava starch biopolymer proved to be efficient in terms of thermal insulation, as it has similar properties to commercially available thermal insulators. Flammability analyses were also carried out and the material was found to be non-flammable. Hardness was also analysed, characterising it as a hard material. As a result, the material proved promising for use as a thermal insulator. These analyses were based on ABNT, ASTM and UL-94 standards and the results obtained experimentally prove that cassava starch biopolymer can be used for thermal insulation purposes.

Key words: Biopolymer, cassava starch, biodegradable, thermal insulator.

SUMMARY

CHAPTER 1

INTRODUCTION

Some natural materials used to be discarded in the environment as rubbish, but with technological advances it has become possible to use this "rubbish" to create products applied to industries. Brazil stands out in this scenario, as it is one of the main countries responsible for research in these areas, mainly because it has a large natural reserve. It has a vast diversity of materials that are little explored and can be used in various applications (SATYANAYANA, 2007).

The environmental degradation caused by industries is noticeable, mainly due to the use of unsuitable materials (DOBIRCAU *et al.*, 2009). For this reason, there is now a stimulus for research into developing technologies based on natural resources that have a high degradation capacity without generating toxic waste. Currently, with these environmental concerns, industries are looking for materials that do not degrade the environment and are low cost. As a result, there has been an increase in studies into the use of plant fibres, which can replace synthetic fibres, as well as natural polymers, which can also replace synthetic polymers (ÁVEROUS; DIGABEL, 2006).

A large proportion of applications that require thermal insulation, especially in domestic, industrial and commercial systems, operate at low and medium temperatures of up to 180°C (YOUNG; FREEDMAN, 2008). As a result, environmentally aggressive materials are generally used, such as glass wool, rock wool, polyurethane, polystyrene, EPS and others. Some of these products are expensive and take years to degrade in nature. For this reason, there is a need to reduce these parameters, thus emphasising the use of natural materials.

Cassava starch, also known as manioc starch, is widely used in the food industry and has been for years. Increasing interest in the development of new materials, especially biodegradable ones, has led to a rise in the number of studies on the use of this raw material as the main component in the creation of new materials. These are most commonly used with foam characteristics, especially in the development of tray-type packaging, which is widely used in the food industry. However, they can be used in a variety of applications, such as matrices for the development of composites, as well as thermal and acoustic insulation.

Biodegradable packaging using cassava starch has advantages over other packaging, especially in terms of low cost and environmental issues. Materials made exclusively from cassava starch have the behaviour of materials used as matrices in the development of composites. For this reason, some reinforcement products have been added in order to increase their mechanical properties, such as plant fibres, which aim to increase these properties, creating renewable and low-cost composites (CHIELLINI *et al.*, 2009; MALI *et al.*, 2010; SATYANARAYANA; ARIZAGA; WYPYCH, 2009).

There are some gaps in this material, as its thermal and mechanical properties have been little explored. The material developed from cassava starch and water has the physical characteristics of foams, which are excellent for thermal insulation. This work is therefore motivated by the possibility of developing an innovative thermal insulator.

1.1   OBJECTIVES

1.1.   1General objective

Developing a biopolymer using cassava starch and water to be used in thermal insulation.

1.1.   2Specific objectives

-Develop   a manufacturing process for the material;

-Analyse   moisture absorption;

-Evaluate   the hardness of the foam when subjected to stress according to its applications;

-Check the loss of mass of the material as the temperature rises, using thermogravimetry (TGA);

-Analyse   the structure of the material using scanning microscopy (SEM);

- Carry out flammability tests to check how flammable the material is;

- Determine the thermophysical properties of the foam, such as: density, thermal conductivity, specific heat, thermal diffusivity, resistivity;

- To compare the properties of conventional thermal insulators available on the market with those of the material developed.

CHAPTER 2

LITERATURE REVIEW

2.1.   BIOPOLYMERS AND THEIR APPLICATIONS

Biopolymers are polymers or copolymers that are developed from raw materials from renewable sources, such as corn, sugar cane, cellulose, chitin and others. Renewable sources have a shorter life cycle compared to sources derived from petroleum, which require a long time to degrade. Some environmental and socio-economic factors that are related to the growing interest in biopolymers are: the major environmental impacts caused by the extraction and refining processes used to produce polymers from oil, the scarcity of oil and the increase in its price (BRITO *etal.*, 2011).

Biodegradable polymers are materials that degrade due to the action of bacteria, fungi and algae. They can be consumed in weeks, months or even years, depending on the environment in which they are used. The materials that have attracted the most attention are those developed from renewable sources, as they have less of an environmental impact. Many biopolymers are produced commercially on an industrial scale, and the main raw materials for their manufacture are renewable carbon sources from carbohydrates derived from large-scale commercial plantations such as sugar cane, corn and cassava (BRITO *etal.*, 2011).

The development of products based on natural polymers, biopolymers and/or biodegradable materials are offering various alternatives in search of a more sustainable development process (Abd El-Rehin, 2006; Ma *el al.*, 2012). In this way, biopolymers are establishing themselves as an important category of materials for reducing the environmental impact of synthetic plastic packaging made from fossil fuels (Mariniello *et al.*, 2003) and the use of different raw materials in the preparation of biodegradable materials is being developed (Liu *etal.*, 2012).

Renewable sources are known as such because they have a shorter life cycle compared to fossil sources such as oil, which takes thousands of years to form. Some environmental and socio-economic factors that are related to the growing interest in biopolymers are: the major environmental impacts caused by the extraction and refining processes used to produce polymers from oil, the scarcity of oil and its rising price. Another preponderant factor is the non-biodegradability of the vast majority of polymers produced from oil, which contributes to the accumulation of plastic waste without an appropriate destination that will take tens to hundreds of years to be assimilated by nature again.

Despite all their advantages, biopolymers have some technical limitations that make them difficult to process and use as a final product. Therefore, many research groups have dedicated themselves to studying the modification of biopolymers in order to enable their processing and use in various applications (FECHINE, 2010). To this end, blends (BALAKRISHNAM *et al.,* 2010; HUNEAULT; LI, 2007), composites (CHEN; WANG, 2002; WONG; SHANKS; HODZIC, 2004), nanocomposites [CHIVRAC *et al.,* 2010; RHIM; HONG; HA, 2009] have been studied in order to improve properties such as processability, thermal resistance, mechanical properties, geological properties, gas permeability and degradation rate.

It is well known that studies into natural polymers are advancing, especially in the development of packaging to replace synthetic polymers, since natural polymers are environmentally friendly and come from renewable and sustainable sources. In this way, it is possible to see that the main aim of replacing fossil waste materials such as polypropylene, polyethylene and other synthetic polymers with natural polymers is to avoid the impact caused by these petroleum-based plastics (KUMAR; SINGH, 2008; KAITH *et al.,* 2010; JOHN; TOMAS, 2008).

Thermoplastic starch can be used in the development of natural materials, called biomaterials, in different applications. Its main features are biodegradability, low cost and properties relatively close to those of synthetic resins (ÁVEROUS, DIGABEL, 2006).

According to John and Thomas (2008), biopolymers can have a long lifespan as long as they are used indoors, i.e. in environments with low humidity.

Starches have generally been used in the development of biopolymers that have low durability applications, but this does not prevent the use of these materials in applications that require longer life. Composite materials made from thermoplastic starch-based polymeric matrices reinforced with plant fibres are therefore responsible for materials with differentiated properties.

Starch is a biodegradable, renewable, inexpensive resource that comes from various plant sources and is produced in abundance. For these reasons, there is great interest in it for non-food purposes, replacing synthetic polymers (ANGELLIER, *etal.,* 2004).

2.2.   MANDICA

According to Embrapa (2017), cassava is known in many regions of Brazil as macaxeira, aipim, castelinha and macamba. Its scientific (botanical) name is *Manihot esculenta Crantz.* It originated in South America and is one of the main foods, especially in developing countries. Cassava is shown in Figure. 1.

Figure 1:   Cassava

Source: EMBRAPA, 2017.

According to Abam (2017), Brazil is the second largest producer of cassava in the world. Brazil is responsible for producing 22.2 million tonnes (11.6% of world production). This shows how important it is for the country to use this raw material beyond the food sector, and to be able to use it to develop new products.

Essentially, cassava roots are made up of a brown skin, peel and pulp. All varieties have their own peculiarities, differing in the amount of components, with the starch content being the most variable in their composition. This variation is due to the type of climate, planting season, cultivation system and time of harvest (CEREDA, 1994; EMBRAPA, 2017). The composition of the roots is shown in Table 1.

Figura 2.   Cassava root composition.

Composition	(%)
Umjdade	62,5
Proteins	0,6
Lipids	0,9
Starch	33,0
Fibres	17
Ashes	1,3

Source: CEREDA, 1994.

On average 60 per cent of all production of these roots goes to human consumption, in second place to animal feed with around 33 per cent, followed by the textile, paper and chemical industries, among others, which consume an average of 7 per cent of this material (PANDEY *etaL,* 2000).

The industrialisation of cassava is seen as an excellent alternative for the valorisation of this material, and can develop the labour market in the area, generating employment and financial returns. The most widely used by-products of cassava are starch, flour and starch (MOORE, 2001).

2.2.1 Obtaining the starch

In the starch extraction process, the first step is to remove the root, wash, peel and grate it so that the granules are released and separated from the fibres and soluble components. After that, they go on to the extraction stage, carried out inside rotating conical sieves which are called GLs, made of steel. Water is then added to these GLs in countercurrent to separate the starch from the fibrous material. The water-laden starch is then

for purification. The starch is concentrated and passed through a rotary vacuum filter to remove the moisture. After this, the material is taken to a dryer in a very short time, where it is dehydrated to moisture contents of between 12 and 13 per cent. Once in powder form, it is cooled, temporarily stored and bagged (LEONEL; CEREDA, 2000; VILELA; FERREIRA, 1987). Figure 3 shows a flowchart of the starch production process.

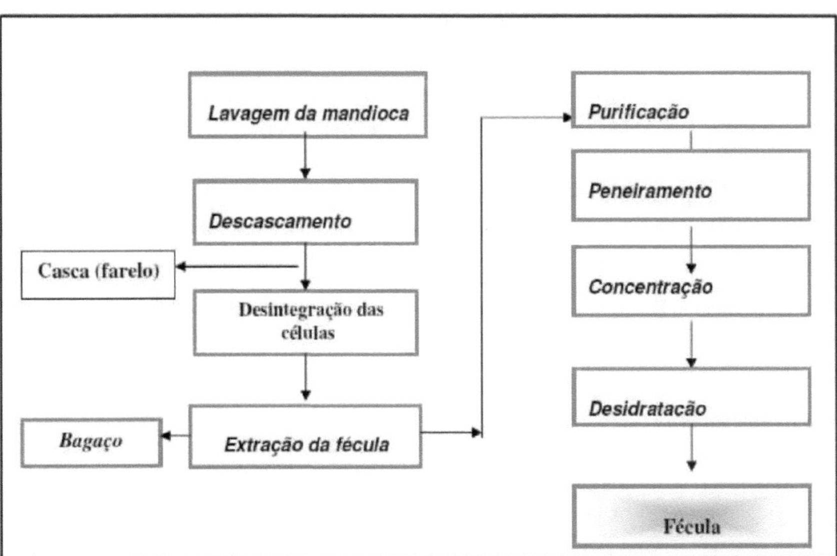

Figura 3. Flowchart of the starch production process.

Source: MATSUI et al., 2004.

1.1.2. Physical and chemical composition of cassava starch

Cassava starch has been studied extensively in the production of biodegradable materials. It

9

can also be known as cassava starch, sweet starch or cassava gum. It is characterised by its fine, white, odourless and tasteless powder. This product can be obtained from cassava roots after peeling, grinding, disintegrating, purifying, sieving, centrifuging, concentrating and drying. Brazil produces this material on a large scale, with Pará being the largest Brazilian producer, contributing around 70.5 per cent of national production (DEBIAGI *etal.*, 2013).

Understanding the structure of starch granules is important for understanding their physicochemical properties, which determine the behaviour of natural or modified starch in the most diverse industrial processes to which they are normally subjected. In terms of the organisation of the granules and the structure of their polymeric constituents, each starch is unique. Groups of plants (cereals, roots, legumes and tubers) and even plants of the same species have starches with different characteristics and properties (RATNAYAKE; JACKSON, 2007).

Starch is a natural polymer (biopolymer) made up of two polysaccharides, amylose and amylopectin, both containing only anhydroglucose units (CeHioOs).

Starch is made up of two types of glucose polymers, amylose and amylopectin, which have different structures. Amylose is a linear polymer, while amylopectin is a highly branched polymer. These components can vary in their composition and properties, resulting in starch grains with different physical, chemical and functional properties, making their industrial applications varied (ELLIS *et al.*, 1998).

Cassava starch has a proportion of amylose between 16-20%, while amylopectin is between 80-84%. This amylose, which are simpler polymers with a linear composition, can create ribbons, fibres or form a mesh that can be very strong and resilient. Amylopectins, on the other hand, are branched polymers with long chains, which can be further classified by the way they branch off from the main chain. Polymers with many branches are known as dendrimers, molecules that can form networks when cooled. This can make the polymer strong in an ideal temperature range. However, when heated, both linear and branched polymers soften when the temperature variation exceeds the attractive forces between the molecules (VICTÓRIA; GROSSMANN; YAMASHITA, 2010).

When creating a starch-based thermoplastic material, its semi-crystalline granular structure needs to be destroyed in order to become a homogeneous and essentially amorphous polymeric matrix. The destruction of the organisation of starch grains is caused by gelatinisation and melting. Gelatinisation is an irreversible transformation of the starch into a viscoelastic paste, which occurs in the presence of excess water, leading to a distribution of the crystallinity and molecular order of the grains due to the breaking of hydrogen bonds. Conversely, when the material is heated with a lower concentration of water, this phenomenon is known as melting, as it requires higher temperatures to occur (VAN SOEST; VLIEGENTHART, 1997; LIU, 2005).

2.3. STATE OF THE ART

2.3.2. Mechanical and thermal properties of cassava starch

Cassava starch has been used industrially in the production of foams (expanded), films, bags and personal hygiene items. This material has mainly been used in the development of biodegradable packaging, with the aim of replacing synthetic packaging with products that are environmentally friendly (BRITO *etal.*, 2011).

According to Parra *et al.* (2004), some elements were added to a biodegradable plastic film using cassava starch and water in order to improve its mechanical properties and water absorption. The plastic film developed with only cassava starch and water obtained results of approximately 0.1 Mpa for tensile strength and 11% for deformation. These properties are directly influenced by the thickness of the material, which is why it proved to have these properties, which are relatively low.

Composites were developed using cassava starch foam as the matrix, which is obtained by extrusion and thermo-pressing, with volume fractions of 80% starch, 19% water and 1% plasticiser (glycerine). In addition, some natural fibres were used as reinforcement in order to comparatively analyse the influence of the composite's properties as the fibre volume fraction increased. The material that did not incorporate fibres (cassava starch foam) had a higher compressive strength of 35 N and flexibility of approximately 5 mm. These properties of this material are extremely important in the development of composites, as it can be used as a matrix (NAIME *etal.*, 2012).

Carr *et al.* (2006) developed a cassava starch foam and then analysed the influence of the quantities of fibres on its properties. The mechanical properties of cassava foam without added fibres were as follows: it can withstand a compressive force of 21 N, a flexibility of 6 mm and a density of 0.2 g/cm^3 , approximately. These values show that the mechanical properties of cassava starch can be used in various applications, specifically in the manufacture of biodegradable packaging.

A composite was developed using cassava starch as the matrix and coconut shell fibre as the reinforcement, showing that increasing the volume of fibre is directly related to an increase in the material's mechanical strength. In addition, studies of the mechanical properties of the cassava starch matrix alone were carried out, which yielded typical curves for matrices (with and without heat treatment), noting that cassava starch has large deformations and low strength (RAMÍREZ *et al.*, 2011).

Also according to Ramízez *et al.* (2011), it can be seen that the matrix-only composite, i.e. without added reinforcement, heat-treated material showed tensile strength, Young's modulus and maximum load of 3.24 Mpa, 59.81 Mpa and 112.68 N, respectively. The non-heat-treated material had a tensile strength of 1.56 Mpa, 14.56 Mpa and 50.77 N, respectively. These values for cassava starch are important because they show that when natural fibres are added, their properties tend to increase. The mechanical properties of these materials are shown in Fig. 4.

These reinforcements are important for increasing the material's mechanical resistance, since without them the material is ductile. It can be seen in Figure 4 that with the inclusion of fibres (from 5%) the deformation of the material drops, remaining approximately constant, making it behave like a rigid material. This demonstrates the feasibility of using cassava starch as a matrix in the development of low-cost, biodegradable composites.

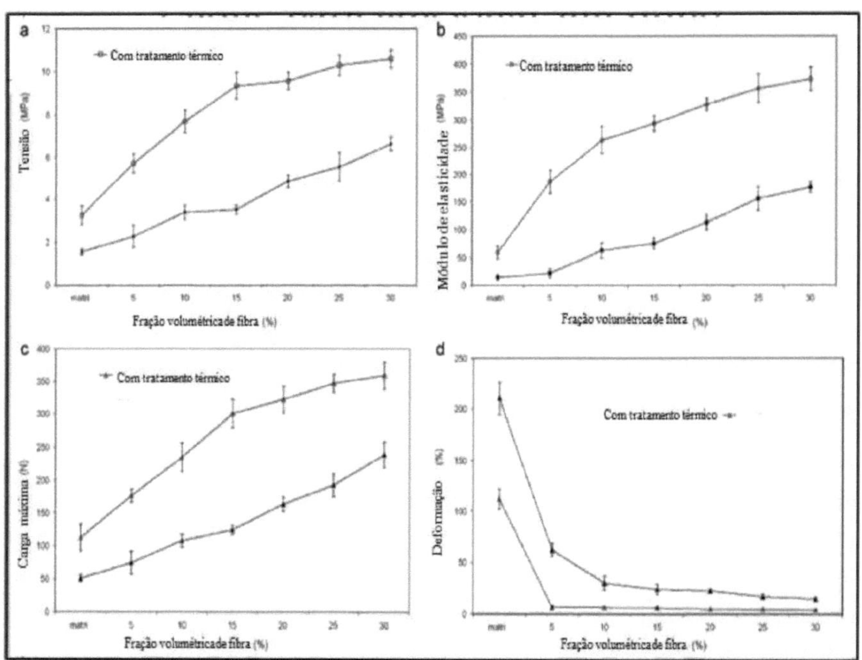

Figure 4: Mechanical properties of the cassava starch and coconut shell fibre composite: (a) stress curve *versus* fibre volume fraction (%). (b) modulus of elasticity *versus* % of fibre, (c) maximum load *versus* % of fibre, (d) deformation *versus* % of fibre.
Source: RAMÍREZ *etal.*, 2011.

Ramírez *et al.* (2014) carried out a thermogravimetric analysis (TGA) on the cassava starch and coconut shell fibre composite, which is shown in Figure 5. The study shows that the variation in the volume fraction of the constituents has no direct influence on this result, because regardless of the amount of fibre, the composite begins to degrade at 300°C (degradation temperature). These results are extremely important as they guide the use of this material in a range of applications, including thermal insulation. However, with some restrictions, especially in high temperature situations where it is not efficient, it can be used on occasions where the temperature remains below 300°C.

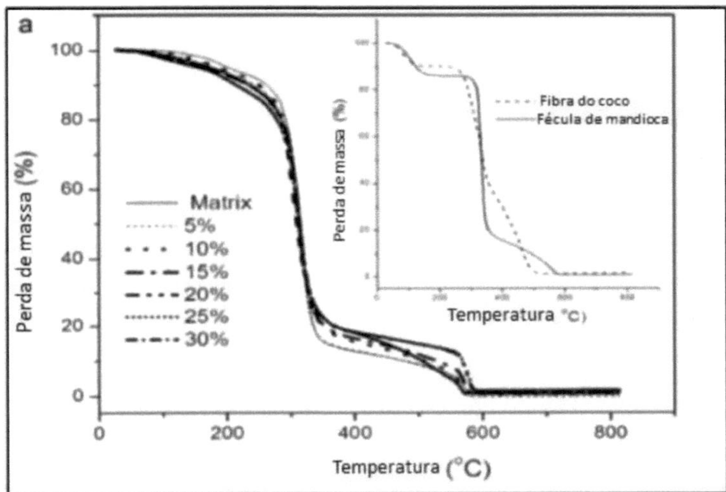

Figure 5 Thermogravimetric analysis (TGA) of the composite using cassava starch as the matrix and coconut shell fibre as the reinforcement.

Source: Ramírez *etal.*, 2014

Figure 5 shows the TGA of the material, which is a destructive technique in the field of thermal analyses, in which the variation in the mass of a sample as a function of temperature or time is monitored in a controlled temperature and atmosphere environment. Its principle of operation is simple: analyse the loss or addition of mass to the sample at varying temperatures. In this way, it can be seen that the material can withstand a temperature of up to 300°C.

Table 1 shows the mechanical properties of cassava starch. These values mean that this material can be used as a matrix together with a reinforcement in the development of a composite, causing the tensile strength of the matrix, cassava starch, to increase with the inclusion of fibres. In this way, cassava starch is shown to be a lightweight material with mechanical properties similar to those of products that are used as matrices. It can therefore be applied in a variety of situations, from packaging to materials that can withstand relatively high loads.

Table 1. Mechanical properties of materials.

Materials	Density (g/cm)³	Tensile Strength (Mpa)	Modulus of elasticity (Mpa)	Stretching (%)
Cassava starch	0,2	1,46-3,24	14,56-59,81	-

Source: RAMÍREZ *et al.*, 2011; RAMÍREZ *etal*, 2014; CARR *etal.*, 2006.

2.3.3. Cassava starch applications

Polymeric materials are considered indispensable in modern life, but due to the development of new technologies, it is increasingly difficult to find a polymer that fully meets all the needs for a given application, whether due to mechanical properties, characteristics, ways of obtaining them or cost. For this reason, polymers are increasingly used as matrices in the development of composites, or even used in a particular application as thermal and acoustic insulation.

The use of pure starch as a material is not recommended, as it is very brittle and sensitive to water, so it is necessary to carry out heat treatments to give the material greater strength, flexibility and water resistance, which are necessary for commercial foam applications. Water resistance can be improved with a polyester coating and natural fibres can be used to increase the mechanical properties of the starch (CARR *etal.*, 2006).

According to Curvelo, Carvalho and Agnelli (2001), Eucalyptus fibres were used as a reinforcement for thermoplastic starch, with the result that the composite developed showed a 100% increase in tensile strength and more than 50% in modulus of elasticity compared to the non-reinforced thermoplastic.

According to Lawton, Shogren and Tiefenbacher (2004), a starch foam composite with 2.5-45% fibre content was developed, and the results showed that the strength of the foam increased as the fibre content increased until it reached around 20%.

Stoffel *et al.* (2015) developed cassava starch foam trays by means of thermal expansion, shown in Figure 6. Their morphology, physicochemical and mechanical properties were analysed. The difference between trays with and without microcellulose was also analysed. In terms of density, the two obtained practically the same results (on average 0.25 g/cm^3). While water absorption was shown to be proportional to time and after 60 minutes both materials had the same absorption, around 300%. The mechanical properties were also not significantly influenced by the addition of 1% microcellulose, obtaining the following average results: tensile strength of 0.61 Mpa and elongation of 0.45%.

Figure 6 - Cassava starch trays.

Source: STOFFEL *et al.*, 2015.

Ramírez *et al.* (2011) developed composites using cassava starch as a matrix and green coconut fibre as reinforcement, which were prepared with different amounts of coconut fibre, up to 30%. The influence of the volume fraction of the fibre on its mechanical properties and water absorption was then analysed. The mechanical properties of the starch were improved with the inclusion of fibres and heat treatments. The results showed that tensile strength, modulus of elasticity and maximum supported loads increased with the amount of fibres used, but elongation decreased. The water absorption, moisture absorption and linear expansion of the composite tended to decrease as the coconut fibre content increased.

Santos *et al.* (2013) developed films with the aim of replacing packaging made from polyethylene. Cassava starch, double-distilled glycerine, hydrogen peroxide and water were used to make the material. The manufacturing process used was to press the material mixture and then leave it to dry in an air circulation oven. It was found that the plasticiser is fundamental to the mixture, as it determines the viscoelastic behaviour of the material, producing a product with a good appearance.

2.4. THEORETICAL BACKGROUND

2.4.2. Thermal insulation

The search for suitable energy-saving methods is an important concern for scientists in relation to human existence. A major contribution to the efficient use of energy and consequent reduction in cost is to insulate thermal systems (TSAI, *et al.*, 2007). With this, parts of thermal systems must be coated with material that has properties and thicknesses such that the temperature remains within a certain range (TORREIRA, 1980).

Thermal insulators are very diverse and can be applied to industrial pipework that transports hot or cold fluids, and to furnace walls. The aim of this procedure is generally to avoid the risk of burns and the heating of environments neighbouring these structures, as well as to reduce heat transfer. The application of insulation to walls is also often used to prevent thermal insalubrity or to avoid energy loss in order to maintain the thermal comfort of the environment (COUTINHO, 2005).

The thermal conductivity of some insulating materials varies depending on their porous structure, whose small cavities hold gases of low thermal conductivity confined in the closed cells. The main factor affecting the thermal conductivity of thermal insulators is the thickness of the material; other factors to consider include the specific mass and size of the material's cells, humidity and ambient temperature (MENDES, 2002).

With wide application in engineering, thermal insulators are selected according to aspects ranging from economic to functional and even safety. The main function of a thermal insulator is to reduce the rate of heat transfer between a system and a medium, so that energy can be conserved (INCROPERA; DEWITT, 2008).

A material's ability to slow down the flow of heat between two media is the main parameter that characterises it as a thermal insulating material. However, in order for it to be classified, a more detailed study of its main mechanical and thermal properties is necessary (KREITH, 2008).

According to Torreira (1980), the properties that a material must possess to be considered a good thermal insulator are: low thermal conductivity, high specific heat, low thermal diffusivity, low specific mass, chemical and physical stability, ease of application, low moisture absorption, absence of odour, low cost, economy and ease of application.

In practice it can be seen that there are no ideal thermal insulators, but it is possible to minimise losses and achieve the insulation you are aiming for. Depending on the application, different types of insulation can be determined according to requirements.

When choosing an insulating material suitable for a particular application, it is important to check various properties of the material to be used, the main ones for thermal insulation being: thermal conductivity, specific heat, thermal diffusivity and its resistance to the stress to which it will be subjected.

2.4.3. Types of thermal insulation

Nowadays there are a vast number of materials that can be used in thermal insulation, where each insulator has a particular application, its use depending on the temperature gradient desired and the climate in which it will be used.

2.4.3.1. Natural insulation

There are many natural or basically natural insulators, which can be of mineral origin (glass wool, rock wool, cellular glass, expanded clay, among others), vegetable origin (expanded cork agglomerates, cellulose fibres, wood fibre agglomerates, coconut, sisal) or even animal origin, such as sheep's wool. These natural materials have environmental advantages over synthetic products derived from petroleum.

Natural fibres are being used as thermal insulators because they have structures composed of empty spaces, which reduces the material's thermal conductivity, as well as being inexpensive. This improves its use as a thermal insulator (NEIRA; MARINHO, 2005).

Various materials can also be used as an insulator in their *natural* form, without undergoing any treatment, because they have a low thermal conductivity value, such as fibres which are classified as: seed fibres (cotton), leaf fibres (sisal, curauá), stem fibres (joint, flax), fruit fibres (coconut) and root fibres (zacatão) (BORGES, 2009).

One of the main advantages of using natural materials for thermal insulation is their low cost compared to synthetic materials. A composite using latex as a matrix and coconut fibre as a reinforcement was developed with the aim of using it for thermal insulation applications in small and medium temperature ranges. As a result, the thermal properties of the composite were quantified and it can be classified as a thermal insulator. Thus, coconut fibre proves to be an excellent thermal insulator (MENDES, *etal.,* 2012).

Natural fibres therefore have enormous advantages over synthetic fibres, such as: their low density and abrasiveness, low energy consumption and costs, biodegradability, recyclability, high specific mechanical resistance properties and excellent thermo-acoustic properties.

2.4.3.2. Synthetic insulation

Synthetic insulators are classified as thermoplastics (plastics), thermosets, rubbers and fibres. Thermoplastics are heat-mouldable and have low density conductivity, are impact resistant, low cost and are good as thermal and electrical insulators (BORGES, 2009).

According to Borges (2009), expanded polystyrene (EPS), also known as Styrofoam, is the most widely used synthetic thermal insulator. It is used in refrigeration systems and heat exchangers. Most of it has been used in civil construction, added to concrete, slabs and other construction systems linked to this sector, with the main aim of improving the thermal performance of structures, insulating floors, ceilings and walls.

2.4.4. Heat transfer

When there is a difference in temperature between regions, or when two bodies with different temperatures are brought into contact, a phenomenon called energy transfer takes place. When a material easily allows the transfer of energy by temperature difference, it is called a thermal conductor, and when a material hinders this transmission, it is called a thermal insulator (YOUNG; FREEDMAN, 2008).

According to Incropera and Dewitt (2008), thermal heat flow is the rate of heat transfer per unit area perpendicular to the direction of transfer and it is proportional to the temperature gradient in this direction. The flow can occur in two ways: stationary thermal flow and transient thermal flow.

Also according to Incropera and Dewitt (2008), which describes the first law of thermodynamics (law of conservation of energy) as the increase in the amount of energy accumulated (stored) in a control volume must be equal to the amount of energy entering the control volume minus the amount of energy leaving the control volume.

The definitions of stationary regime and transient regime are in terms of heat flow. When the flow inside the wall is constant, the regime is stationary (permanent), i.e. the incoming flow is the same as the outgoing flow. In the transient regime there is a variation in the heat flow in the different sections of the wall, so the incoming flow is different from the outgoing flow (INCROPERA; DEWITT, 2008).

Steady-state conduction is the form of conduction that occurs when the temperature difference is constant so that after an equilibrium time, the spatial distribution of temperatures in the conducting object does not change. For the transient case, on the other hand, the heat flow and the temperatures inside the objects change over time, until a new steady state of thermal gradient is reached, or when a state of uniform temperature is reached (which always happens, eventually, if the external temperatures and the heat sources and their sinks are not altered).

It is possible to quantify the heat transfer process in terms of the equation for heat flow by conduction. This equation is known as Fourier's law.

2.4.4   Fourier's Law

Fourier's law is a phenomenological law, i.e. a law developed from observed phenomena and not deduced from fundamental principles. We therefore regard the equation for heat flow by conduction as a generalisation based on a large body of experimental evidence (INCROPERA; DEWITT, 2008).

According to Incropera and Dewitt (2008), this law, also known as the law of thermal conduction, shows that the rate of heat transfer through a material is proportional to the negative temperature gradient and the cross-sectional area of the gradient through which the heat is flowing, as shown in Equation 1.

$$q = -kA\frac{dT}{dx} \qquad (1)$$

When the above expression applies to the one-dimensional case, when there is a temperature gradient in the x direction. Once the surface temperatures are known, the above equation is integrated to give the following equation:

$$q = \frac{kA\ \Delta T}{L} \qquad (2)$$

Where:

-q = Heat rate (W);

-L = Wall thickness (m);

- ΔT = Temperature change in the interval considered (°C);

-k = Proportionality constant called thermal conductivity (W /m°C);

-The   ratio $\frac{\Delta T}{\Delta x}$ - is called the temperature gradient;

The negative sign is due to the heat being transferred in the direction of decreasing temperatures (INCROPERA; DEWITT, 2008).

2.4.5 Thermal conductivity

Thermal conductivity is considered the most important of all the thermal properties of an insulator, which defines the ability of a material to conduct a greater or lesser amount of heat. It can be expressed by the equation Eq. 1 (INCROPERA; DEWITT, 2008).

Most thermal insulators are porous in nature, including solid matter and voids. The voids contain gases that have low thermal conductivity, leaving the insulator with a lower conductivity value. The conditions that can influence this property are: humidity, the orientation of the insulator fibre in relation to the direction of flow (KREITH, 2008).Figure 7 shows various ranges of thermal conductivity of various states of matter at normal temperatures and pressures.

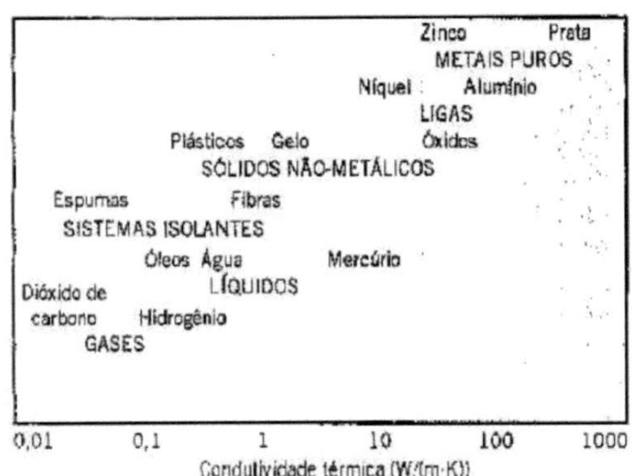

Figure 7. Thermal conductivity of materials.

Source: DEWITT & INCROPERA, 2008.

Thermal conductivity is also affected by the density of the material. For each type of insulator, fibrous or cellular, there is a minimum conductivity value for a particular density above which the conductivity increases if the density is increased or decreased. In the case of fibrous insulators, a

A decrease in fibre diameter reduces conductivity for the same density. Low-density fibre materials allow a high percentage of heat transmission by radiation and show a greater variation in thermal conductivity as a function of changes in temperature and emissivity of the surrounding surfaces (KREITH, 2008).

According to Incropera and Dewitt (2008), humidity is a critical condition when the temperature of the insulation is lower than the dew point temperature of the surrounding air, causing condensation of the atmospheric water vapour on the insulation. The effect of humidity on the insulation's conductivity is not so worrying while it is in the vapour phase. However, conductivity is affected and increased by the presence of condensed moisture in the material.

There is an analogy between the diffusion of heat and electrical charge, as electrical resistance is associated with electrical conduction. Thermal resistance is associated with heat conduction. Defining thermal resistance for conduction in a flat wall as: "the ratio between a driving percentage and the corresponding rate of transfer". (INCROPERA; DEWITT, 2008), it is defined by Equation 3.

$$R_{condução} = \frac{L}{kA} \tag{3}$$

Thermal conductivity values are provided by specialised literature as a function of density, geometry and average working temperature. These "k" values are quantified by specific equipment in accordance with the ASTM-C177 standard (1997).

20

2.4.6. Specific heat

According to Çengel and Boles (2006), specific heat is defined as: "the energy required to raise the temperature of a unit mass of a substance by one degree". Both specific heat at pressure and at constant volume represent the same definition. It can be measured using a calorimeter.

For liquids and solids, the values of cp and cv are approximately the same. They only vary for gaseous fluids. The specific heat in terms of enthalpy is shown in Equation 4.

$$c_p = \frac{\partial h}{\partial T} \tag{4}$$

Where:

-c_p = Specific heat at constant pressure (J/kg.K);

-h = Specific enthalpy (J/kg);

-T = Temperature (K);

Or defined in terms of internal energy, by Equation 5.

$$c_v = \frac{\partial u}{\partial T} \tag{5}$$

Where:

-c_v = Specific heat at constant volume (J/kg.K);

-u = Internal energy of the system (J/kg);

The thermal capacity of a material (C) is defined by the ratio of heat supplied to or removed from the material and the temperature variation. It is expressed by Eq. 6 (INCROPERA; DEWITT, 2008).

$$C = \frac{dQ}{dT} \tag{6}$$

Where:

-dQ$\quad$= Energy required to produce a change in temperature (J);

-dT$\quad$= Temperature change (K);

-C$\quad$= heat capacity (J/K);

Even though the energy of the material increases with temperature, the amount needed to produce a one-degree change in temperature is the same.

2.4.7   Thermal diffusivity

The ratio between thermal conductivity and volumetric heat capacity is an important property called thermal diffusivity, which is defined by Equation 7 (INCROPERA; DEWITT, 2008).

$$\alpha = \frac{k}{\rho c p} \tag{7}$$

Where:

-k = thermal conductivity (W/mK);

-p = specific mass (kg/m$)^3$;

   $-c_p$ = specific heat (J/kgK);

- α = thermal diffusivity (m^2 /s);

It is a property that characterises the amount of thermal energy it can conduct in relation to its capacity to absorb it. The higher the material's diffusivity, the less time it takes to reach thermal equilibrium (INCROPERA; DEWITT, 2008).

According to Santos *et al.* (2004), thermal diffusivity is a measure of how heat propagates in matter. It is an important property in all problems involving steady-state heat conduction.

Thermal diffusivity is an important variable for thermal control, as it shows how quickly the body inside adjusts to the surrounding temperature. Materials with low thermal diffusivity slow down heat transfer, meaning that it takes longer to reach the permanent regime.

CHAPTER 3

METHODOLOGY

3.1. BIOPOLYMER MANUFACTURE

The raw materials used to make the biopolymer were cassava starch and water. The cassava starch (Figure 8a) was purchased at street markets in the city of Natal/RN, while the water used was distilled at the Fluid Mechanics Laboratory - NTI/UFRN. In addition to these materials, it was necessary to use an anti-moulding agent (Figure 8b) in order to make it easier to remove the material from the mould, which was purchased from the Hidro Glass shop located in Natal/RN.

(a)   (b)

Figure 8. Materials used in manufacturing: (a) cassava starch; (b) anti-moulding agent.

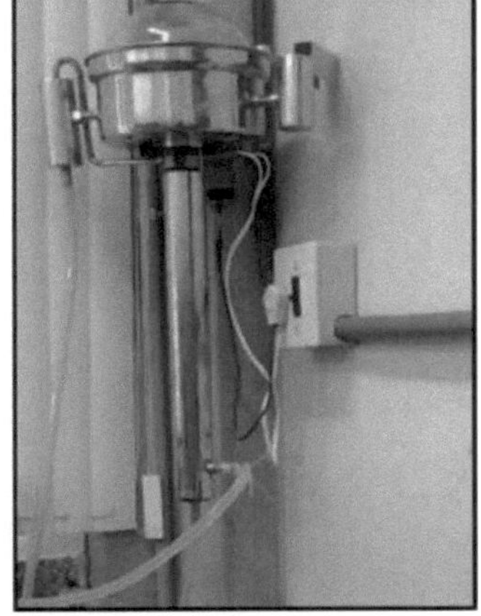

Figure 9. Distiller

In order to avoid impurities that could interfere with the properties of the biopolymer, distilled water was used, which was acquired through a distiller, shown in Figure 9.

The manufacturing process defined for the development of the cassava starch biopolymer (cassava starch foam) is thermal expansion, which is carried out as follows: the two materials (starch and water) are mixed manually in a beaker and then poured into the mould. After this, it is taken to the oven at a controlled temperature so that the material expands spontaneously, leaving it in the form of a natural foam.

However, in order to develop the material, it was realised that some variables needed to be controlled, such as: volume fractions of the constituents, temperature and time spent in the furnace. In view of this, it was necessary to try and vary these parameters in order to arrive at the ideal ones.

It was therefore necessary to vary the volume fractions of the starch from 90 to 70 per cent, the water from 10 to 30 per cent, the oven temperature from 150 to 300°C and the time spent in the oven from 5 to 90 minutes. These variations were fundamental to finding the reference variables at which the material could obtain a satisfactory result.

The ideal proportion found for the material was 70 per cent starch and 30 per cent water, as well as a temperature of 250°C. These variables produced the best results, making it possible to develop a well-distributed, compact and light foam, as shown in Figure 10. While the materials with lower proportions of water were difficult to handle due to the high viscosity of the mixture, as well as having low expansion, those with higher proportions had a large number of voids and very thin walls. These problems would make it difficult to quantify their properties. After manufacturing the cassava starch and water biopolymer in an open mould, it was possible to see that it had an undesirable void in its centre, as shown in Figure 10. It was therefore necessary to develop closed moulds in order to contain this expansion, which would avoid this void in the centre.

Figura 10.   Foam with 70% starch and 30% water.

For this reason, the experiments were carried out in a closed 304 stainless steel mould, designed in SOLIDWORK and manufactured in the energy laboratory - NTI / UFRN. The mould design is shown in Figure 11a, while the actual mould is shown in Figure 11b.

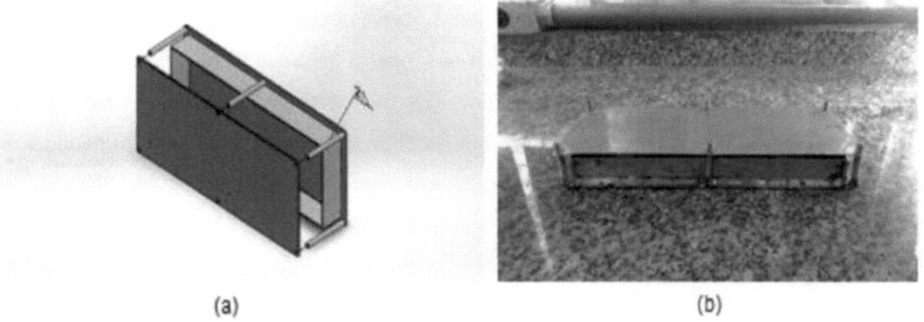

(a)   (b)

Figura 11.   Mould: (a) drawing; (b) real.

In addition to this mould, we also used a cylindrical mould with threads on the bottom and studs on the top (Figure 12a). Once again, the result was unsatisfactory, as the material expands in these environments and the internal pressure rises, causing it to leak through the gaps, which is why, when removed from the mould, the material showed a void in its centre, mainly due to its leakage (Figure 12b).

(a)   (b)

Figura 12.   Closed mould manufacturing: (a) cylindrical mould; (b) material with a void in the centre.

Faced with this problem, the method used to manufacture the material was expansion control. The mould used for this purpose was made of aluminium, with dimensions of 200x100x50 millimetres, in order to facilitate the removal of the material together with the anti-moulding agent.

The expansion control method consists of placing the material in the oven, removing it after 30 minutes and inserting a steel plate (with a mass of 1.9kg) into its surface. The material is then placed back in the oven for the required time. In this way, the void in the centre of the material can be avoided.

By changing the moulds used to make the material, it was necessary to vary the length of time it remained in the oven, due to the variation in its mass, mainly because the moulds had different dimensions. As a result, it was essential to check how long the material took to expand. 60 minutes was used, but the material didn't expand completely, as shown in Figure 13a. After 90 minutes, it was noticeable that the material expanded completely and evenly, as shown in Figure 13b.

Figura 13.   Cassava starch biopolymer: (a) incomplete expansion; (b) complete expansion

In this way, it was possible to define all the variables needed to manufacture the cassava starch biopolymer. The parameters for producing the material were: 500 ml (70 per cent) of cassava starch, 150 ml (30 per cent) of distilled water, an expansion temperature of 250°C and a time spent in the oven of 90 min;

The new material was developed by manually mixing cassava starch and water in a beaker. The mixture was then poured into a mould and placed in an oven at a controlled temperature for 90 minutes. The oven used to make the biopolymer is from the manufacturer Zezimaq, model 2000, with a working temperature limit of up to 1300°C with variations of 1°C, seen in Figure 14a, while Figure 14b shows the material taken to the oven to be expanded.

Figura 14.   Material manufacture: (a) used furnace; (b) material in the furnace.

After the time spent in the oven, the material undergoes controlled expansion. After this procedure, the cassava starch foam is cooled in the oven for 24 hours in order to avoid possible thermal shocks and to homogenise its structure. Therefore, at the end of the process, the biopolymer has become a porous, light and rigid foam, which is shown in Figure 15.

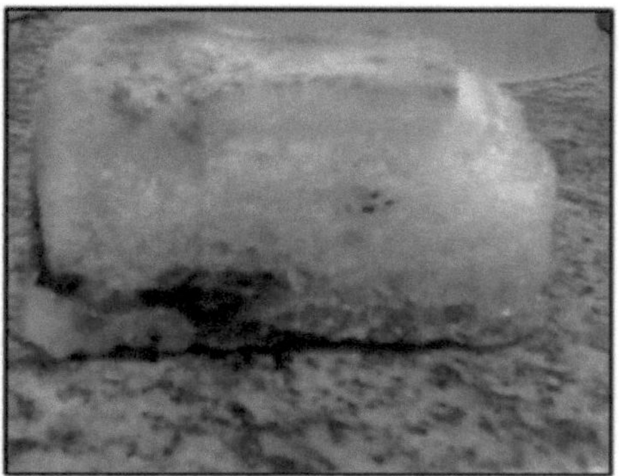

Figure 15 - Cassava starch biopolymer.

3.2   DETERMINING THE SPECIFIC MASS

The tests to determine the specific mass (density) of the cassava starch biopolymer were carried out on a Gehaka model DSL 910 digital densimeter, which uses the Archimedes principle (thrust) and is suitable for solids and liquids. This equipment is available at the Fluid Mechanics Laboratory - NTI / UFRN.

These tests used 5 samples of the material with dimensions of 30x30x30 millimetres (Figure 18a), in accordance with the requirements of ASTM D 792 (2008).

To carry out the test, the samples were cut to standard dimensions. The densimeter is then calibrated using a beaker with distilled water (water density 1.05g/cm^3) and a temperature sensor. The sample is then taken to the weighing pan, where the sample is weighed (dry mass), as shown in Figure 18b. The sample is then submerged in the water contained in the beaker using a convex basket (wet mass), as shown in Figure 18c. Given this data, using Archimedes' principle, the balance gives the result of the material's density in g/cm^3 .

27

(a)   (b)   (c)

Figure 16. Specific mass of the material: (a) samples; (b) weighing on the plate; (c) submerged sample.

3.3.   MOISTURE ABSORPTION ANALYSIS

The aim of determining the percentage of moisture contained in the cassava starch foam was to check the capacity of this material to absorb moisture from the ambient air into which it was inserted. To do this, a Marte ID-200 humidity scale was used, with sensitivity (mass) of 0.01 g and precision (humidity) of 0.1%. This is available at the NTI/UFRN Fluid Mechanics Laboratory. Figure 17 shows the humidity determiner used.

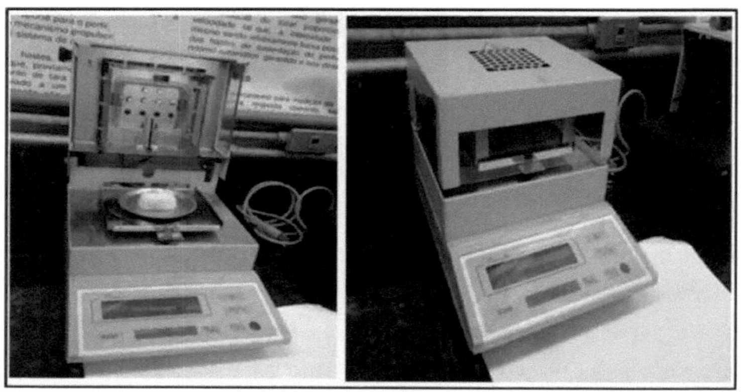

Figure 17 - Mars ID-200 scale.

The Mars ID-200 (Series Interface Detectors) humidity determiner has an infrared heat source produced by a quartz encapsulated resistor, temperature adjustment from 60 °C to 180 °C, with a 1 °C increment, humidity range: 0-100%. Bi-directional RS232C interface, LCD (Liquid Crystal Display) illuminated dot matrix display.

The procedure required to carry out the experiment on this equipment was done in accordance with the manufacturer's manual, which states that the sample of material analysed must have a minimum initial mass of 1 gram and a maximum of 200 grams.

This work used 5 samples of the biopolymer with standard dimensions of 30x20x20 millimetres (Figure 18a). These samples were exposed to the environment for 7 days during which their temperatures and relative humidity (maximum and minimum) were quantified. The equipment used to measure these environmental parameters during these days is shown in Figure 18b.

(a)   (b)

Figure 18 - Moisture absorption: (a) samples; (b) equipment used to measure relative humidity.

Each sample was inserted into the balance's heating unit, where it was exposed to a temperature of 110 °C until it reached its constant mass. This temperature was used to ensure that the water contained in the material evaporated. The time needed for the mass to remain constant was 40 minutes. After this time, values were obtained for the initial mass, final mass, percentage of moisture removed, working temperature and duration of the experiment.

3.4. DETERMINATION OF SPECIFIC HEAT, CONDUCTIVITY AND THERMAL DIFFUSIVITY

These tests make it possible to determine the material's thermal properties. The KD2 Pro equipment (Figure 19a), available at the Fluid Mechanics Laboratory - NTI / UFRN, was used for this purpose. It has an SH-1 sensor (double thermal needles) which is used to capture values after 2 minutes of insertion into the test specimens, this sensor works in the range of 0.02 to 2.00 W/m-K.

Holes were drilled into the material using a bench drill with a 2-inch diameter bit. These holes were drilled in pairs, with the necessary spacing for the equipment's sensors to penetrate the material, as seen in Figure 19b. These needles (sensors) have a length of 30 mm and a diameter of 1.27 mm and the equipment basically works by creating a flow of heat between these needles. The tests were carried out in accordance with the equipment manufacturer's manual. Four specimens measuring 120x7x5 millimetres were analysed and three measurements were taken at different points for each one. Figure 19b shows the test being carried out.

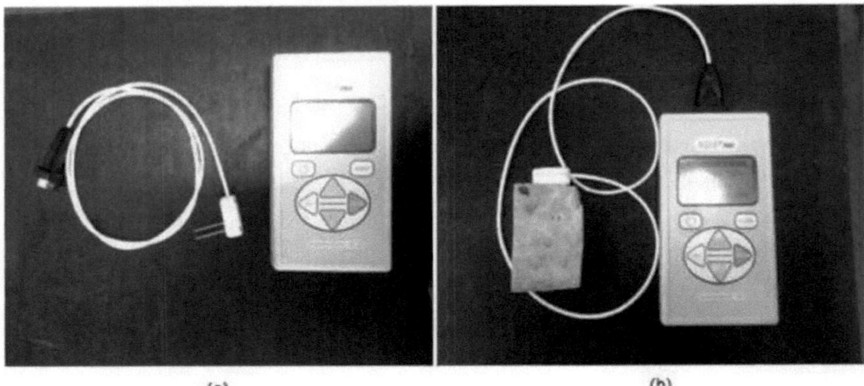

(a)   (b)

Figure 19. Thermal analysis: (a) KD2 PRO equipment; (b) test.

3.5.   SHORE D HARDNESS TEST

The hardness of a material expresses its resistance to permanent deformation, which can be referred to as the ability of one material to penetrate another. This property is directly related to the bonding strength of the material's atoms. In addition, the term hardness can also be associated with resistance to bending, scratching, abrasion or cutting.

The hardness tests in this work were carried out at the Tribology and Dynamics Laboratory - NTI/UFRN, using a Shore D Portable Analogue Durometer from Kori Seki MFG. LTD (Japan), shown in Figure 20a. Co. LTD (Japan), shown in Figure 20a. According to the ASTM D 2240 standard (2003), this equipment is indicated for measuring the hardness of rubbers, plastics and foams that are rigid, which is measured by pressing a steel penetrator into the test specimen. In this test, 4 specimens with dimensions of 100x50x30 millimetres were used. 4 measurements were taken at different points on each specimen. Figure 20b shows how the test was carried out.

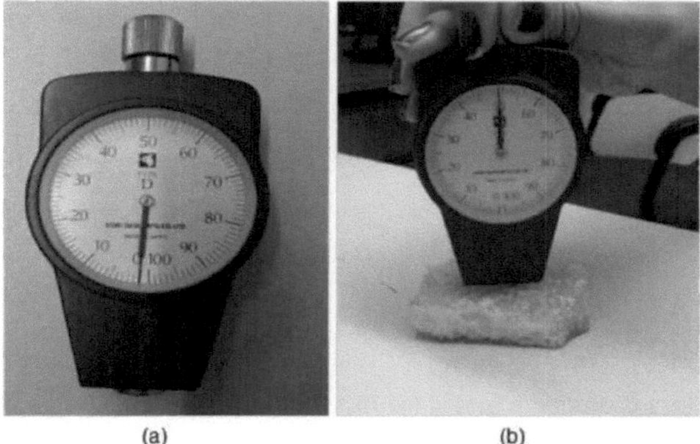
(a)   (b)

Figure 20. Hardness test: (a) Shore D durometer; (b) test in progress.

3.6. THERMOGRAVIMETRIC ANALYSIS - TGA

Thermogravimetry (TG) is the technique in which the mass of a substance is measured as a function of temperature, while the substance is subjected to a controlled temperature programme. This technique basically consists of

a high-precision balance associated with an oven, in which you can control the heating rate (usually 10 °C/min) or keep the temperature constant. The atmosphere to which the sample is subjected can also be controlled.

In this work, thermogravimetric analyses were carried out on 2 samples of cassava starch biopolymer, which had a mass of 10mg. The analyses were carried out using the NETZSCH TG209F1 Libra simultaneous thermogravimetric and calorimetric analyser, available in the Thermal Analysis laboratory at the Institute of Chemistry/UFRN. In accordance with the ASTM D6370 standard, the samples were placed in an alumina crucible and subjected to temperatures ranging from ambient to 700 °C, at a heating rate of 5 °C/min in an air atmosphere, with a flow rate of 20 ml/min.

3.7. SCANNING ELECTRON MICROSCOPY - MEV

The scanning electron microscopy (SEM) technique allows samples to be observed many times over with good image resolution, providing morphological characterisation of the material being analysed.

The principle of operation of the SEM consists of the incidence of a high-energy electron beam on the surface of the sample where an interaction occurs and part of this beam is reflected and collected by the collector. Images are usually obtained by secondary electrons (SE), which come from inelastic interactions between the electrons and the sample, or they can be acquired by backscattered electrons (BSE), which come from elastic interactions between the electrons. The images obtained by secondary electrons are more commonly used in SEM, as they provide a higher resolution image with a three-dimensional impression and are easy to interpret.

To carry out the test, 3 samples of cassava starch biopolymer were used, each measuring 30x20x20 millimetres. The main aim of this analysis was to check the porosity and morphological characteristics of the material.

The analyses were carried out on a HITACHI model TM3000 scanning electron microscope (SEM), as shown in Figure 21a. The samples were sent to the DEMAT/UFRN Materials Structural Characterisation Laboratory, where they were fixed to the metal sample holder using a carbon strip. The samples were then placed in the equipment, as shown in Figure 21b, and the analyses were carried out.

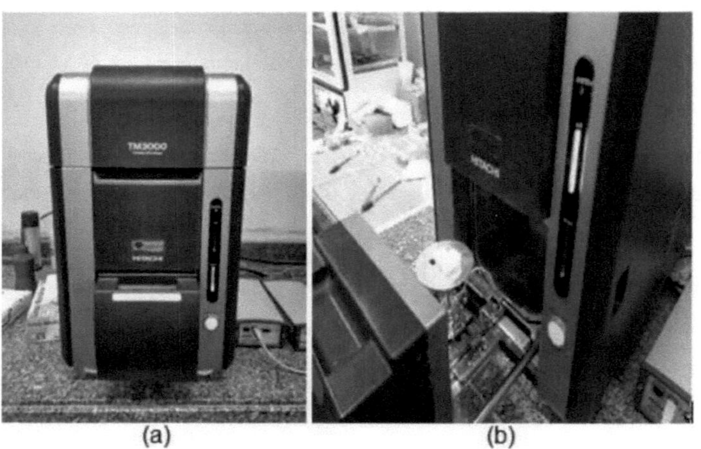
(a)   (b)

Figure 21. SEM: (a) Equipment; (b) Analysis.

3.8. FLAMMABILITY TEST

Among the most widely used methods for measuring flammability parameters, the UL94 (2001) standards are the most widely used internationally for determining these parameters. UL94 is a standard imposed by the quality control laboratory of Underwriters Laboratories (UL) that examines the flammability and fire behaviour of polymer materials.

In this work, the flammability tests were carried out in accordance with the UL94 standard, which recommends that the tests be carried out using the UL94 HB (Horizontal Burning) or UL94 V (Vertical Burning) method. The method used was UL94 HB, in which the object is tested in a horizontal position in front of the flame.

This UL94 HB test consists of igniting a flame in a specimen with average dimensions of 125 mm in length, 13 mm in width and a minimum thickness of 3 mm and a maximum of 13 mm.

This standard is used in order to classify cassava starch biopolymer as UL94 HB through the burning speed of the material. In this test, carried out at the NTI/UFRN Fluid Mechanics Laboratory, 4 samples were used, two of which were flame retardant and the other two were not. The aim of this test is to check whether the material is flammable and to ascertain the influence of the flame retardant on the flammability of the material.

This test used samples 125 mm long, 13 mm wide and 3 mm thick. The samples were marked 6 mm, 25 mm and 100 mm from the end to be burnt. The samples were attached to the hook by the unmarked end, with their longitudinal axis placed in a horizontal position. The Bunsen burner should be inclined at an angle of 45° and adjusted to produce a blue flame equal to 20 mm in length in the vertical position. Figure 22 shows the flammability test as it is carried out.

Figure 22. Flammability test

The flame should be applied to the free end of the specimen, 6 mm mark, for a period of 30 seconds. If the specimen continues to burn after the flame has been removed, record the time taken for combustion to reach the 25 mm mark and, if combustion continues, record the time taken for combustion to reach the 100 mm mark. If the 100 mm mark is not reached, record the time and the length damaged.

The HB classification is awarded if the following conditions are met:

-The   burning speed is no more than 38 mm/min for specimens with thicknesses between 3 and 13 mm.

- The burning speed is no greater than 75 mm/min for specimens with a thickness of less than 3 mm.
- The fire must extinguish itself before 100 mm is reached.

The purpose of the UL94 horizontal flammability test (UL94 HB) is to determine the linear burning rate (V), in millimetres per minute, for each specimen, using Eq. 8.

$$V = 60\frac{L}{t} \tag{8}$$

In which:

V = Linear firing rate (mm/min);

L = Damaged length of the specimen (mm);

t = Time (s).

CHAPTER 4

RESULTS AND DISCUSSION

4.1.   SPECIFIC MASS

Table 3 shows the density values of the 5 samples. The average result for the samples was 0.6692 g/cm^3 , i.e. 669.20 kg/m^3 . This characterises it as a light material.

Table 2. Material density values.

Samples	Density (g/cm)^3	Temperature (°C)
1	0,679	25
2	0,678	25
3	0,634	25
4	0,654	25
5	0,701	25

Knowledge of the specific mass of materials is of fundamental importance, as it influences the total weight of the system to which the material is applied. As a comparative parameter, Figure 23 shows the density of some commercially available materials used in thermal insulation.

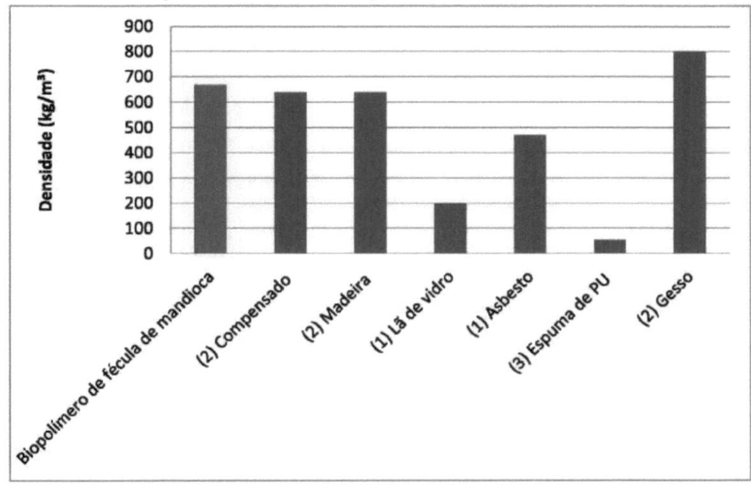

Figure 23. Comparison with some thermal insulators.

Source: (1) KREITH, 2008; (2) INCROPERA; DEWITT, 2008; (3) GALVÃO, 2014.

The biopolymer produced has the characteristics of a foam, but compared to polyurethane foam, it has a higher density. However, its density is close to that of wood and lower than that of plaster, both of which are considered thermal insulating materials. In this way, the biopolymer developed is relatively light and has a density within the range of materials considered insulating, facilitating its use in appropriate applications.

4.2. MOISTURE ABSORPTION

The data from the 7 days during which the samples were exposed to the environment was quantified and is shown in Table 3. This analysis shows that the maximum and minimum relative humidity levels during the week were 84.7% and 61.7%, respectively.

Table 3. Environmental data during the week.

Days	Maximum temperature (°C)	Minimum temperature (°C)	Maximum relative humidity (%)	Minimum relative humidity (%)
Monday	29,4	21,8	83	75
Tuesday	33	23,1	84	68
Wednesday	33,6	22,9	87	64
Thursday	35,3	22,6	86	58
Friday	35,5	23,7	86	5S
Saturday	35,7	23	84	60
Sunday	35,3	23,3	83	48

The moisture absorption results of the samples can be seen in Table 4. The average amount of moisture absorbed by the material after 7 days in the environment is 9.92%.

Table 4. Moisture contained in the material.

Samples	1	2	3	4	5
Humidity (%)	9.7	9.6	9.5	10,2	10.6

Table 5 shows a comparative analysis of the moisture absorption of some insulating materials available on the market. It can be seen that cassava starch biopolymer has moisture absorption values close to that of wood and approximately 3 times higher than polyurethane foam.

Table 5. Moisture absorption of some thermal insulators.

Materials	Biopolymer	Polyurethane (1)	Glass wool (1)	Wood (2)
Moisture absorption (%)	9,92	3,258	0,124	12 à 18

Source: (1) KREITH, 2008; (2) ABNT, 1997.

The moisture contained in the material is a negative factor for thermal insulators, as it induces an increase in thermal conductivity. This relatively high absorption is due to the porosity (empty spaces) which facilitates this infiltration.

This moisture in the material can be reduced by using a waterproofing agent. This high moisture absorption is characteristic of foams in general, which are lightweight and can be used for thermal and acoustic insulation.

Analysing Table 5, it can be seen that the moisture absorption of the biopolymer is lower than that of wood, which is used in thermal insulation for walls, ceilings, windows and so on. As such, the material can be used in similar applications to wood and can be used in the construction industry.

4.3. THERMOPHYSICAL PROPERTIES

Three tests were carried out on each specimen. Table 6 shows the average of the 3 tests for each sample. The average values of the material's thermal properties are: thermal conductivity 0.152 W/mK, thermal resistivity 686.529 °C.cm/W, specific heat 1.210 MJ/m³ K, thermal diffusivity 0.125 mm² /s.

Table 6. Thermal properties of the material.

Specimens				
	1	2	3	4
k(W/mK)	0,1397	0,151	0,174	0,145
R fC.cm/W)	719,117	704,6	629,65	692,75
cp (MJ/m³ K)	1,331	1,307	1,177	1,084
α(mm² /s)	0,106	0,114	0,146	0,134
T("C)	27,043	28,15	'27,245	27,065
Error	0,00143	0,00105	0,00285	0,0015

4.3.1 Thermal conductivity

Thermal conductivity is one of the most important properties of thermal insulators; its average value is 0.152 W/mK for natural polymer (biopolymer). Figure 24 shows a comparison between the conductivities of insulating materials.

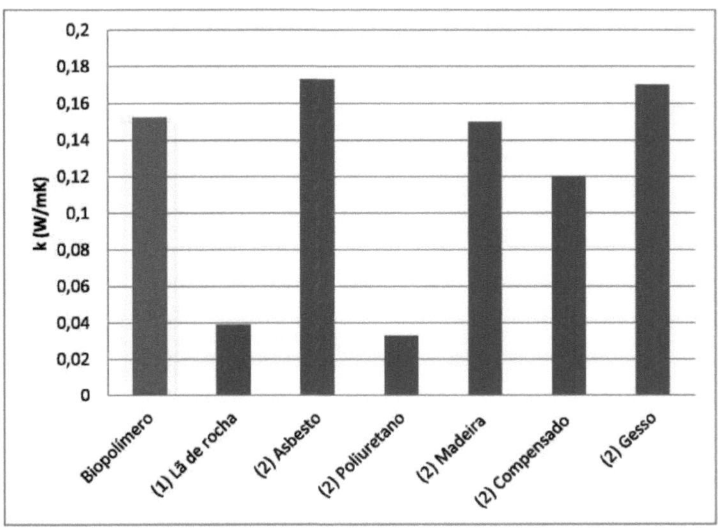

Figure 24. Thermal conductivity of insulating materials.

With regard to the thermal conductivity results, the lower the value obtained, the better the material's thermal insulation properties. A comparative analysis of the thermal insulating materials shown in Figure 24 shows that cassava starch biopolymer has thermal conductivity values close to those of asbestos, wood, plywood and plaster. This means that it has a thermal conductivity value that classifies it as a thermal insulator.

Cassava starch biopolymer has a thermal conductivity value similar to materials used mainly in construction, which makes it viable for use in similar applications. In view of this, it can be seen that cassava starch biopolymer has a wide range of applications in thermal insulation, as it is a material with low thermal conductivity and, above all, because it is a biodegradable material.

4.3.2 Specific heat

The material has an average specific heat value of 1.210 MJ/m^3 K. Figure 25 shows a comparative analysis of the specific heat values of some materials used in thermal insulation.

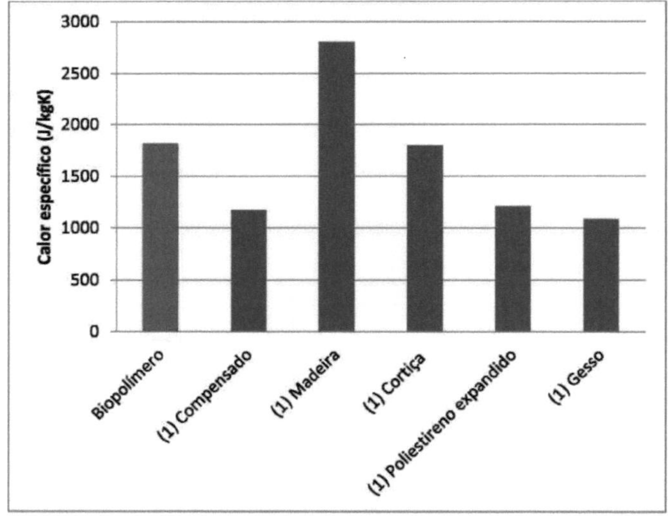

Figura 25. Specific heats of insulating materials.

The equipment quantifies the specific heat of the material as a function of volume (MJ/m^3 K), the capacity to store energy in a cubic metre. The conventional values found in tables are expressed in mass (J/kgK), the capacity to store energy in one kilogram (INCROPERA; DEWITT, 2008). It is therefore necessary to convert the specific heat of the biopolymer into the same unit as the materials found in the literature in order to carry out a comparative analysis between them. Therefore, the value of the specific heat must be divided by the density of the material (666.92 kg/m^3), so the value of the specific heat of cassava foam can be expressed as 1814.31 J/kgK.

The results showed that the material has a specific heat close to that of cork. In view of this, it is possible that the natural polymer has similar applications to cork, which is used in thermal insulation. Therefore, the cassava starch biopolymer has a specific heat value close to that of commercially available thermal insulators, which classifies it as an insulating material.

Like wood, cork has various applications, including cork stoppers and cork flooring, which are the two most representative products in the industry. However, this material has been increasingly exploited by architects and engineers from the most varied sectors of the world, who have used this raw material to insulate walls and ceilings, among other things. Cassava starch biopolymer can be used in similar applications due to its ability to absorb heat.

25.3.3. Thermal diffusivity

The material has an average thermal diffusivity value of 0.125 mm^2 /s. Figure 26 shows the values of this property for some materials that are considered thermal insulators.

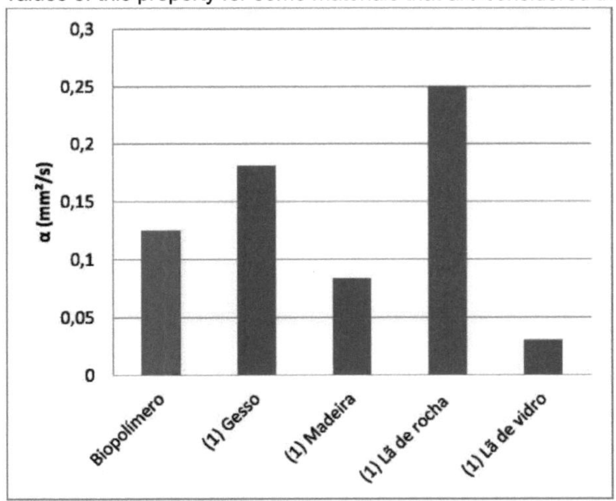

Figura 26.   Thermal diffusivity of insulating materials.

Source: INCROPERA; DEWITT, 2008.

These low values express well (as defined by the thermal diffusivity equation) the biopolymer's low heat transfer capacity and high energy storage capacity, enabling it to be used as a thermal insulating material. In this way, it is possible that this new material could have applications similar to those of products with similar thermal properties (wood, asbestos and plaster).

4.4. SHORE D HARDNESS

Table 7 shows the average results for each specimen. As a result, the cassava starch biopolymer has an average hardness result of 57.5 ShD.

Table 7. Shore D hardness test results.

Test body	1	2	3	4
Shore D hardness	57,5	62,5	53,75	56,25

Figure 27 shows a Shore D hardness graph of some materials that are used as thermal insulators.

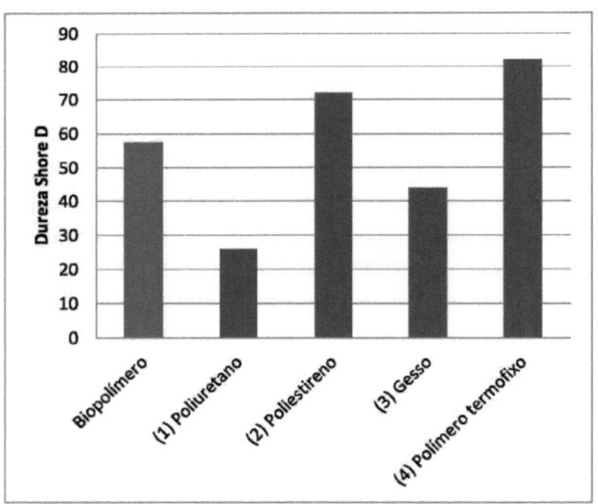

Figure 27. Comparative hardness values of some thermal insulators.
Source: (1) CÂMARA, 2015; (2) COUTINHO, 2007; (3) MOPIC, 2017; (4) PAIVA, 1999.

A comparative analysis shows that the cassava starch biopolymer has a hardness value between polystyrene and gypsum. When the material is compared with polyurethane foam, it can be seen that it has a hardness value approximately twice that of polyurethane.

This difference in the hardness of cassava starch foam and that of polyurethane (PU) characterises the biopolymer as a more rigid foam than PU foam. This is due to the material's low tenacity, i.e. fragility when subjected to medium impacts. The cassava starch biopolymer is therefore a hard material.

Polystyrene has a variety of applications and can be used to make disposable cups and cutlery, as well as being used in expanded form (Styrofoam) for thermal insulation. Gypsum, meanwhile, is widely used in construction for thermal insulation of ceilings and walls.

With similar hardness to these materials, cassava starch biopolymer can be used in similar applications. Thus, this material can be used in the manufacture of flat plates or pipe troughs, which are used for thermal insulation.

4.5.   THERMOGRAVIMETRIC ANALYSIS - TGA

Figure 28 and Figure 29 show the TG/DTG curves of the two cassava starch biopolymer samples. The curves show the mass losses in two stages and the thermal events corresponding to these losses. The two samples show similar TG/DTG profiles, confirming the reliability of the test data.

Analysing the curves shows that there is a mass loss of approximately 11% due to dehydration associated with an endothermic peak around 100 °C. The percentage of weight loss at this stage depends on the moisture content of the material. According to LIU *et al.* (2009), the initial temperature of starch degradation does not affect the start of the decomposition temperature, because all the water evaporates first.

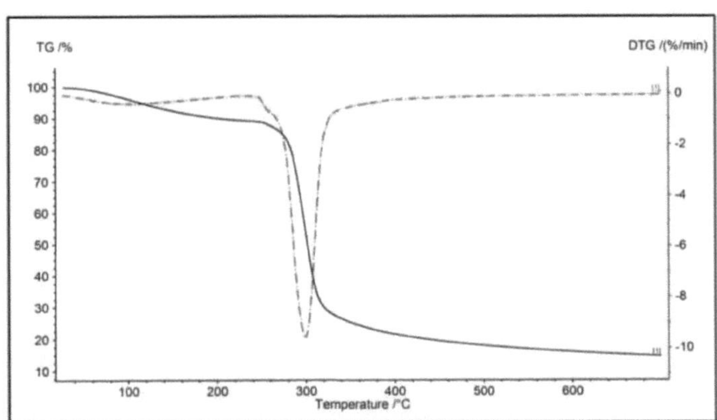

Figura 28.   TD/DTG curves for sample 1.

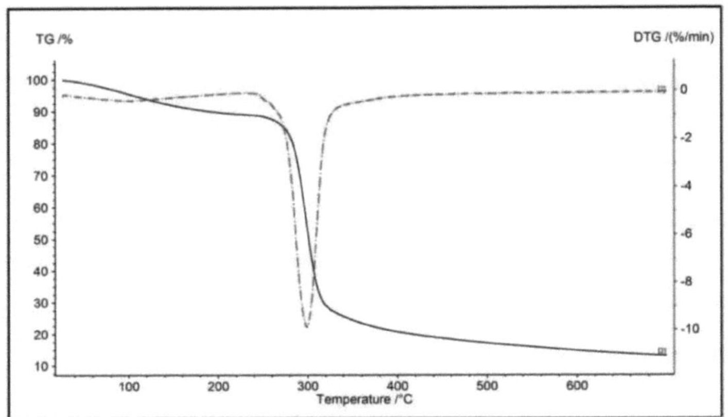

Figura 29.   TG/DTG curve of sample 2.

Thermal decomposition of the cassava starch biopolymer begins after its dehydration. According to Fig. 28 and 29, the total mass loss (%) for the material was approximately 88%. The temperature at which thermal degradation began was approximately 270 °C, while the DTG peak value was 300 °C.

The results obtained through thermogravimetric analyses (TGA) showed that the cassava starch biopolymer can be used in hot systems, with a temperature limit of up to 270 °C, since the material does not degrade up to this temperature.

This temperature limit covers the majority of low and medium temperature thermal applications used in the domestic, industrial and commercial spheres (YOUNG; FREEDMAN, 2008).

1.6.   SCANNING ELECTRON MICROSCOPY - MEV

Three samples of cassava starch biopolymer were analysed using scanning electron microscopy (SEM) at 200 and 2000 times each. The images from these analyses can detect defects in the structure, such as micro-cracks and impurities. In addition, the presence and dimensions of voids and pores can be verified, which influence the material's ability to retain heat.

Figure 30 shows the morphology of sample 1 of the cassava starch biopolymer obtained by scanning electron microscopy (SEM) at 200x and 2000x magnification. The micrographs (a) and (b) show the formation of closed pores that are clusters of voids, which make the material less resistant due to tension contractions between the voids. On the other hand, these voids have a direct influence on the material's thermal conductivity, as they contain confined air which reduces the material's thermal conductivity.

(a)   (b)

Figure 30. Microscopy of sample 1: (a) 200x; (b) 2000x.

It can be seen in Figure 30 that there is a variation in the dimensions of the pores in the material and that the concentrations of these pores are random, for example, there are regions with high pore density and others with low, characterising the material as non-uniform.

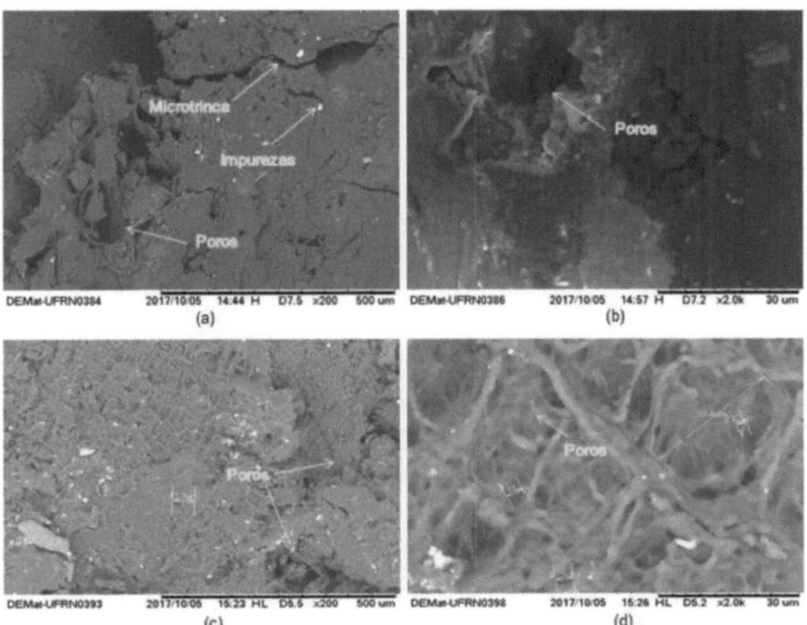

(c)   (d)

Figure 31. SEM: (a) 200x sample 2; (b) 2000x sample 2; (c) 200x sample 3; (d)2000x sample 3.

Figure 31 shows the scanning electron microscopy (SEM) of samples 2 and 3 of the cassava starch biopolymer. These sample quantities are important to check that the morphological characteristics of the material are maintained.

43

It can also be seen in Figure 31 that the material has pores of different sizes at different points, confirming that there is no uniform distribution of these pores. As a result, the material may show points that are more efficient in terms of thermal insulation than others. However, as seen in the thermal conductivity test, there is a slight dispersion of these values, but it is not significant.

Figure 31a shows the presence of impurities on the surface of the material, which are usually a result of the manufacturing process and/or the cutting of the samples. Also in Figure 31b, the biopolymer showed small cracks, which were not visible to the naked eye. These cracks are due to the cutting of the samples, because when subjected to concentrations of tension, due to the cutting of the samples, cracks can appear.

The cassava starch biopolymer can therefore be characterised as a light, porous and hard material. As a result, it can be used in a variety of thermal applications, as it has the characteristics of a thermal insulator. One of its applications as a thermal insulator can be thermal insulating boards, used in construction for insulating walls, ceilings and windows, or it can be used as a thermal insulating pipe rail, widely used in vapour transmission and air conditioning systems.

1.7.   FLAMMABILITY

Four specimens were analysed, two with flame retardant and two without. Figure 32 shows the test being carried out.

(a)   (b)

Figure 32. Flammability test: (a) contact with flame, (b) burning

Once the flame time had elapsed on the specimens, it was realised that the material did not self-combust, i.e. when the flame lost contact with the material, it was extinguished in the sample. As a result, the burning time of the material was short and the damaged length of the sample was practically zero, as shown in Table 8. The specimens that used flame retardant were called specimens 1R and 2R, while those that didn't were called specimens 1 and 2.

Table 8. Flammability test data.

Samples	Damaged length (mm)	Time (s)
1	0	2
2	0	3
1R	0	1
2R	0	1

Analysing the results, it can be seen that the burning speed (flame speed) is practically zero, i.e. the material does not self-combust. Figure 33 shows the burning of the material, which only charred the surface area where the flame came into direct contact with the material. As such, there was no deterioration of the material, showing that it does not self-combust, therefore classifying it as non-flammable.

Figure 33. Result of the flammability test

As can be seen in Figure 33, there was no degradation of the material. The centre of the cassava starch biopolymer did not burn, showing that the material is not flammable. It can be seen that the use of the retardant does not have a significant influence on the self-combustion of the material, because with its use, only the surface carbonisation was lower.

It was observed that even during the flame sustaining time, all the specimens analysed did not emit smoke or soot, nor did they drip dark liquid. These are characteristics of petroleum-based polymers. Thus, cassava starch biopolymer is an excellent material that is not polluting and is biodegradable. It therefore has thermal properties similar to commercially available thermal insulators. The material is therefore viable for thermal insulation applications.

As a result, it can be concluded that cassava starch biopolymer can be classified according to the UL-94 HB standard, which prescribes that the maximum firing speed should not exceed 38 mm/min for specimens with thicknesses between 3 and 13 mm.

The damaged length of the sample is very small, meaning you would need a measuring instrument with high precision to make this measurement. Therefore, the value of this length is approximately zero. However, even with precise instruments, the burning speed would be very low, close to zero, so the material can still be classified according to the UL94 HB standard.

4.8. SUMMARY OF PROPERTIES (AVERAGE VALUE)

Table 9. Summary of properties

K (w/mK)	a (mm² /s)	Cp (J/kgK)	TGA ro	Hardness (ShD)	Density (kg/m)³	Moisture absorption [%]	Flammability
0.152	0.125	1814,31	270	57.5	EG9.2	9.92	Non-inflammatory

CHAPTER 5

CONCLUSIONS

1. According to the thermal parameters analysed, it can be said that cassava starch biopolymer can be used as a thermal insulator, since the values found for these parameters are compatible with those of conventional thermal insulators. It can be concluded that cassava starch biopolymer is an innovative thermal insulator, proving to be efficient in retaining heat.
2. In terms of specific mass, it can be concluded that the cassava starch biopolymer has values close to those of commercially available thermal insulators, especially wood and plaster. The material is therefore characterised as light and can be used in a wide range of applications.
3. The cassava starch biopolymer had a moisture absorption of 9.92%. It can be concluded that this moisture absorption value is due to the large number of pores, which are characteristic of foams. Therefore, the material is classified as a cassava starch foam.
4. In terms of Shore D hardness, the cassava starch biopolymer was found to have a hardness of 57.5 ShD. This value classifies the material as hard.
5. Thermogravimetric analysis shows that it can be used in hot systems of up to 270 °C, contributing to environmental issues as it is a biodegradable material.
6. Scanning electron microscopy showed a very porous material with numerous cavities and a repeating structure.
7. In terms of flammability, based on the results obtained, it can be said that cassava starch biopolymer is a non-flammable material.

SUGGESTIONS FOR FUTURE WORK

The research could be extended to check the influence of other parameters not covered at this stage of the work, such as

1. Quantify the mechanical properties of cassava starch biopolymer;
2. Check the influence of adding fillers (reinforcement) to the biopolymer;
3. Developing ways for large-scale production.

REFERENCES

ABAM - BRAZILIAN ASSOCIATION OF CASSAVA STARCH PRODUCERS. Available at: <http: www.aban.orq.br>. Accessed on 30 July 2017.

ABD EL-REHIN, H. A. Characterisation and possible agricultural application of polycrylamide/sodium alginate crosslinked hydrogels prepared by ionizing radiation. J. AppL Polym. Sei., v. 101, n. 6, p.

3572-3580, 2006.

ABNT. Design of timber structures. ABNT / NBR 7190. Rio de Janeiro, 1997.

ANGELLIER, H.; CHOISNARD, L.; BOISSEAU, M. S.; OZIL, P.; DUFRESNE, A. Optimisation of the preparation of aqueous suspensions of waxy maize starch nanocrystals using a response surfasse methodology. Biomacromolecules. V. 5. P. 1545-1551, 2004.

ASTM C177. Standard Test Method for Steady-State Heat Flux Measurements and Thermal Transmission Properties by Means of the Guarded- Hot-Plate Apparatus, 1997.

ASTM D 2240. Standard Test Methoud for Rubber Property - Durometer Hardness. 2003.

ASTM D 792. Standard Methouds for Density and Specific Gravity (Relative Density) of plastics by Displacement. 2008.

AVÉROUS, L.; BOQUILLON, N. Biocomposites based on plasticised starch: thermal and mechanical behevirours. Carbohydrate polymers. V. 59, p. 111-122, 2004.

AVÉROUS, L.; DIGABEL, F. L. Properties of biocomposites based on lignocellulosic fillers. Carbohydrate polymers. V. 66, p. 480-493, 2006.

BALAKRISHNAN, H.; HASSAN, A.; WAHIT, M. U.; YUSSUF, A. A.; RAZAK, S. B. A. Novel toughened polylactic acid nanocomposite: Mechanical, thermal and morphological properties. *Materials & Design.* V. 31, n. 7, 2010.

BORGES, J. C. S. Composite of castor bean polyurethane and vermiculite for thermal insulation. Master's dissertation. Postgraduate programme in mechanical engineering, Federal University of Rio Grande do Norte, Natal, 2009.

BRITO, G. F.; AGRAWAL, P.; ARAÚJO, E. M.; MÉLO, T. J. A. Biopolymers, biodegradable polymers and green polymers. Electronic Journal of Materials and Processes, v.6.2, 127-139, 2011.

CÂMARA, J. R. L. Obtaining and characterising a composite based on castor bean polyurethane and roof tile waste for use as a thermal insulator. Master's thesis from the Postgraduate Programme in Mechanical Engineering (PPGEM) - Federal University of Rio Grande do Norte - UFRN, 2015.

CARR, L. G.; PARRA, D. F.; PONCE, P.; LUGÃO, A. B.; BUCHLER, P. M. Influence of fibres on the mechanical properties of cassava starch foams. Journal of Polymers and the Environment, v. 14, n. 2, p. 179-183, 2006.

ÇENGEL, Y.; BOLES, M. Thermodynamics. McGraw-Hill, São Paulo, 5 ed. 2006.

CEREDA, M. P. Residues from the industrialisation of cassava in Brazil. São Paulo: Editora Paulicélia, 1994.

CHEN, L. J.; WANG, M. Production and evaluation of biodegradable composites based on PHB-PHV copolymer. *Biomaterials.* V. 23, n. 13, 2002.

CHIELLINI, E.; CINELLI, P.; ILIEVA, V. I.; IMAM, S.H.; LAWTON, J.L.; Environmentally compatible foamed articles based on potato starch, corn fibre, and poly(vinyl alcohol). J. CelL Plast. 45, p.17-32, 2009.

CHIVRAC, F.; POLLET, E.; DOLE, P.; Avérous, L. Starchbased nano-biocomposites: Plasticiser impact on the montmorillonite exfoliation process. *Carbohydrate Polymers.* V. 79, n. 4,2010.

COUTINHO, A.S. Conforto e insalubridade térmica em ambientes de trabalho. João Pessoa: PPGEP Editions, 2005. 2 edition.

COUTINHO, F. M. B.; COSTA, M. P. M.; GUIMARÃES, M. J. O. C.; SOARES, B. G. Comparative Study of Different Types of Polybutadiene in the Tenacification of

Polystyrene. Red de Revistas Científicas de América Latina y el Caribe, Espana y Portugal, 2007.

CURVELO, AAS.; CARVALHO, AJF.; AGNELLI, JAM. Thermoplastic starch- cellulosic fibres composites: preliminary results. Carbohydrate Polymers. 45, p. 183- 188, 2001.

DEBIAGI, F.; IVANO, L. R. P.F.M.; NASCIMENTO, P. H. A.; MALI, S. Biodegradable starch packaging reinforced with lignocellulosic fibres from agro-industrial waste. BBR - Biochemistry and Biotechnology Reports, v. 1, n. 2, p. 57-67, 2013.

DORBICAU, L.; SREEKUMAR, P. A; SAIAH, R.,; LEBLANC, N.; TERRIER, C.; GATTIN, R; SAITER, J. M. Wheat flour thermoplastic matrix reinforced by waste cotton fibre: Agro-green-composites. Composites: Part A. v. 40, p. 329-334, 2009.

ELLIS, R.P.; COCHRANE, M. P.; DALE, M. F. B.; DUFFUS, C. M.; LYNN, A.; MORRISON, I. M.; PRENTICE, R. D. M.; SWANSTON, J. S.; TILLER, S. A. Starch production and industrial use (Review). Journal of Science Food and Agriculture, London, v.77, n. 3, p.289-311, 1998.

EMBRAPA - BRAZILIAN AGRICULTURAL RESEARCH COMPANY. Available at: <https://www.embrapa.br/mandioca-e-fruticultura>. Accessed on 28 July 2017.

GALVÃO, A. C. P. Obtaining and characterising a composite of castor bean polyurethane and glass powder for thermal insulation applications. Master's thesis from the Postgraduate Programme in

Mechanical Engineering (PPGEM) - Federal University of Rio Grande do Norte - UFRN, 2014.

HUNEAULT, M. A.; LI, H. Morphology and properties of compatibilised polylactide/thermoplastic starch blends. *Polymer.* V. 48, n. 1, 2007.

INCROPERA, F.P.; DEWITT, D. P. Fundamentals of Heat and Mass Transfer. 6ª ed. Rio de Janeiro. Editora LTC. 2008.

KAITH, B. S.; JUNDAL, R.; JANA, A. K.; MAITI, M. Development of corn starch based green composites reinforced with Saccharum spontaneum L. fibre and graft

copolymer - Evaluation of thermal, physico-chemical and machanical properties. Bioresource Technology. V.101, p. 6843-6851,2010.

KREITH. F. Principles of heat transmission. 9. Ed. São Paulo: Edgard Blucher, 2008.

KUMAR, A. P; SINGH, R. P. Biocomposites of cellulose reinforced starch: Improvement of properties by photo-induced-crosslinking. Bioresource Technology. V. 99, p. 8803-8809, 2008.

LAWTON, J. W" SHOGREN, R. L" TIEFENBACHER, K. F. Aspen fibre addition improves the mechanical properties of baked cornstarch foams. *Ind. Crops Prod.,* 19, p. 41-48, 2004.

LEONEL, M.; CEREDA, M. P. Extraction of the starch retained in the fibrous residue of the cassava starch production process. Food Science and Technology. V. 20, n. 1, 2000.

LIU, H.; XIE, F.; YU, L.; CHENA, L.; LI, L. Thermal Processing of starch-based polymers. Progress in Polymer Science. 34: 1348-1368, 2009.

LIU, M.; YIN, Y.; FAN, Z.; ZHENG, X.; SHEN, S.; DENG, P.; ZHENG, C.; TENG, H.; ZHANG, W. The effects of gamma-irradiation on the structure, thermal resistace and mechanical properties of the PLA/EVOH blends. NucL Instr. Met. Phys. Res. B, v. 274, p. 139-144, 2012.

LIU, Z. Edible films and coatings from starches. In: HAN, J. H. (Ed.). *Innovations in foodpackagings,* Elsevier: Amsterdam, The Netherlands, p. 318-336, 2005.

MA, W.; TANG, C. H.; YIN, S. W.; YANG, X. Q.; WANG, Q.. LIU, F.; WEI, Z. H. Characterisation of gelatin-based adible fi lms incorporated with olive oil. Food Res. Int., v. 49, n.1, p. 572-579, 2012.

MALI, S.; DEBIAGI, F.; GROSSMANN, M. V. E.; YAMASHITA, F. Starch, sugarcane bagasse fibre, and polyvinyl alcohol effects on extruded foam properties: A mixture design approach. Industrial Crops and Products, v. 32, p. 353-359, 2010.

MARINIELLO, L.; DI PIERRÔ, P.; ESPOSITO, C.; SORRENTINO, A.; MAIS, P.; PORTA, R. Preparation and machanical properties of adible pectin-soy flour films obtained in the absence or presence of transglutaminase. J. Biotechnology, v. 102, n. 2, p. 191-198, 2003.

MATSUI, K. N.; LAROTONDA, F. D S.; PAES, S. S; LUIZ, D. B; PIRES, A. T. N.; LAURINDO, J. B. Cassava bagasse-Kraft paper composites: Analyses of influence of impregnation with starch acetate on tensile strength and water absorption properties. Carbohydrates Polymer. V. 55, p. 237-243, 2004.

MENDES, J. U. L.; SILVEIRA, F. F.; CAVALCANTE, S. L. L.; OLIVEIRA, L. K. R.; RIBEIRO, F. A.; SOUSA, R. F. Determining the thermal diffusivity of a natural composite in order to classify it as a thermal insulator. 20th CBECIMAT, 2012.

MENDES, L. U. L. Development of a biodegradable composite for thermal insulation. PhD thesis. Postgraduate programme in Materials Science and Engineering at the Federal University of Rio Grande do Norte. Natal- RN, 2002, 141 p.

MOORE, G. R. P. Maize and cassava starch in the production of maltodextrins. 2001. 85p. Master's dissertation in food science - Federal University of Santa Catarina, Florianópolis, 2001.

MOPIC - SOCIEDADE COMERCIAL DE ANGOLA, LTDA. Available at < http://www.grupomopic.com/admin/ficheiros/artigosFicheiros/49_12022015041302_M_opic_Gesso Acabamento 02122015.pdf>. Accessed on 15 November 2017.

NAIME, N.; BRANT, A. J. C.; LUGÃO, A. B.; PONCE, P. Cassava starch foams with natural fibres. 20th CBECIMAT, 2012.

NEIRA, D. S. M.; MARINHO, G. S. Thermal insulation with post-consumer EPS reduces energy costs. Revista Plástico Industrial. Year VII, No. 84, p. 122-124, 2005.

PAIVA, J. M. F.; TRINDADE, W. G.; FROLLINE, E. Composites of Phenolic Thermoset Matrix Reinforced with Vegetable Fibres. Polymers: Science and Technology, 1999.

PANDEY, A.; SOCCOL, C.; NIGAN, P; SOCCOL, V. T.; VANDENBERGHE, L. P. S.; MOHAN, R. Biotechnolocigal potential of agro-industrial residue II: cassava bagasse. Bioresource Technology, v. 74, p. 81-87, 2000.

PARRA, D.F.; TADINI, C.C.; PONCE, P.; LUGÃO, A.B. Mechanical properties and water vapour transmission in some blends of cassava starch edible films. Carbohydrate Polymers, v. 58: 475-481,2004.

RAMIREZ, M. G. L.; SATYANARAYANA, K. G.; IWAKIRI, S.; MUNIZ, G. B.; TANOBE, V.; FLORES-SAHAGUN, T. S. Study of the properties of biocomposites. Part I. Cassava starch-green coirfibers from BraziL Carbohydrate Polymers, v. 86, n. 4, p. 1712-1722, 2011.

RAMIREZ, M. G. L.; SATYANARAYANA, K. G.; GONZÁLEZ, R. M.; IWAKIRI, S.; MUNIZ, G. B.; FLORES-SAHAGUN, T. S. Bio-composites of cassava starch-green coconut fibenPart II - Structure and properties. Carbohydrate Polymers, v. 102, p. 576-583, 2014.

RATNAYAKE, W. S.; JACKSON, D. S. A new insight into the gelatinisation process of native starches. Carbohydrate Polymers, v. 67, n. 4, p. 511-529, 2007.

RHIM, J. W.; Hong, S. L; Ha, C. S. Tensile, water vapour barrier and antimicrobial properties of PLA/nanoclay composite films. *LWT - Food Science and Technology.* V. 42, n. 2, 2009.

SANTOS, B.; SILVA, R. S.; COELHO, T. M.; ASSAD, N. F. Production of bioplastic from cassava starch. VIII International Meeting of Scientific Production (VIII EPCC), 2013.

SANTOS, W. N.; GREGÓRIO FILHO, R.; MUMMERY, P.; WALLWORK, A. Hot wire method for determining the thermal properties of polymers. Polymers: Science and Technology 14, n. 5, p. 354-359, 2004.

SATYANAYANA, K. G. Biodegradable composites based on lignocellulosic fibres. Paper presented at: International conference on advanced materials and composites (ICAMC-2007), P. 12-24, 2007.

SATYANARAYANA, K.G.; ARIZAGA, G.G.C.; WYPYCH, F. Biodegradable composites based on lignocellulosic fibres - An overview. Progress in Polymer Science, v. 34, p. 982-1021, 2009.

STOFFEL, F.; WESCHENFELDER, E. F.; PIEMOLINI-BARROS, L. T.; ZENI, M. Preparation and characterisation of cassava starch foam trays with chitosan and microcellulose. Revista Iberoamericana de Polímeros, v. 16, p. 260-266, 2015.

TORREIRA, R.P. Isolamento térmico. São Paulo: Fulton Editora Técnica Ltda, 1980.

TSAI, I. J.; LEI, C. H.; LEE, C. H.; LEE, Y. C.; LOU, C. W.; LIN, J. H. Manufacturing process and property analysis of industrial flame retarded PET fibre and polyurethane composite. Journal of Materials Processing Technology. p. 415-421, 2007.

UL 94. Standard for Tests for Flammability of Plastic Materials for Parts in Devices and Appliances. 2013.

VAN SOEST, J. J. G; VLIEGENTHART, J. F. G. Crystallinity in starch plastics: consequences for material properties. *Trends in Biotechnology,* Kidlington Oxford, v. 15, n. 6, p. 208-213,1997.

VICTÓRIA, M.; GROSSMANN, E.; YAMASHITA, F. Starch films: production , properties and potential of utilization Starch films: production , properties and potential of utilization. p. 137-156, 2010.

VILELA, E. R.; FERREIRA, M. E. Technology for the production and utilisation of cassava starch. Agricultural report. V. 13, n. 145, p. 69-74, 1987.

WONG, S.; SHANKS, R.; HODZIC, A. Interfacial improvements in poly(3- hydroxybutyrate)-flax fibre composites with hydrogen bonding additives. *Composites Science and Technology.* V. 64, n. 9, 2004.

YOUNG, H. D.; FREEDMAN, R. A., PHYSICS IV - OPTICS AND MODERN PHYSICS, 12ª ed. São Paulo, Addison Wesley, 2008.

Buy your books fast and straightforward online - at one of world's fastest growing online book stores! Environmentally sound due to Print-on-Demand technologies.

Buy your books online at
www.morebooks.shop

Kaufen Sie Ihre Bücher schnell und unkompliziert online – auf einer der am schnellsten wachsenden Buchhandelsplattformen weltweit! Dank Print-On-Demand umwelt- und ressourcenschonend produzi ert.

Bücher schneller online kaufen
www.morebooks.shop

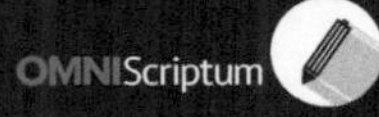

Printed by Books on Demand GmbH, Norderstedt / Germany